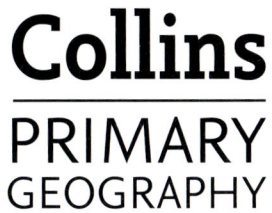

Movement
Pupil Book 4

Planet Earth

Coasts
The seashore — 2
Shaping the coast — 4
Exploring the coast — 6

Water

Rivers
Describing rivers — 8
Rivers matter — 10
Managing rivers — 12

Weather

Weather patterns
Extreme weather — 14
Weather forecasts — 16
Recording the weather — 18

Settlements

Towns
Understanding towns — 20
The origin of towns — 22
Town life — 24

Work and Travel

Food and shops
Farms and food — 26
From farm to supermarket — 28
Local shops — 30

Environment

Caring for towns
Old and new buildings — 32
Making improvements — 34
Comparing places — 36

Places

Northern Ireland — 38
Germany — 44
North America — 50
Asia — 56

Glossary — 62

Index — 63

Stephen Scoffham | Colin Bridge

Unit 1 Coasts

Lesson 1: The seashore

What is the seashore like?

The United Kingdom (the UK) has one of the most varied coastlines in the world. If you went on an aeroplane ride around the coast you would see sandy beaches, mudflats, shingle banks, rocky shores and cliffs.

Discussion
- What are the main types of UK coastline?
- Why do you think some coastlines are straight while others jut out into the sea?
- How might coasts be affected if climate change brings storms and higher sea levels?

▼ A sandy beach in Portrush, Northern Ireland.

Key words
beach
cliff
coast
mudflats
shingle

Unit 1 Coasts

▲ A shingle bank in the Highlands, Scotland.

Data bank
- The coastline of the UK is 12 429 km long.
- No one in the UK lives more than 120 km from the sea.

▲ Mudflats in Leigh-on-Sea, Essex.

▲ Steep chalk cliffs at The Needles, Isle of Wight.

▲ A rocky coastline, Cornwall.

Mapwork
Make a picture map of an imaginary island. Show three or more different coastlines and invent names for them.

Investigation
Write a short description of one of the seashore photographs. Ask someone else to read the description and guess the photograph you have chosen.

Unit 1 Coasts

Lesson 2: Shaping the coast

How does the sea shape the coast?

As waves wash up against the coast they change its shape. In some places the sea wears away the land and makes cliffs and headlands. In other places it builds up beaches in sheltered bays. Usually these changes happen very slowly and take hundreds of years. However, very rough seas can break down part of a cliff face in just a few hours.

Key words
bay
boulder
cave
cliff
headland
rock stack
sand dune

▼ Sand dunes, Portugal.

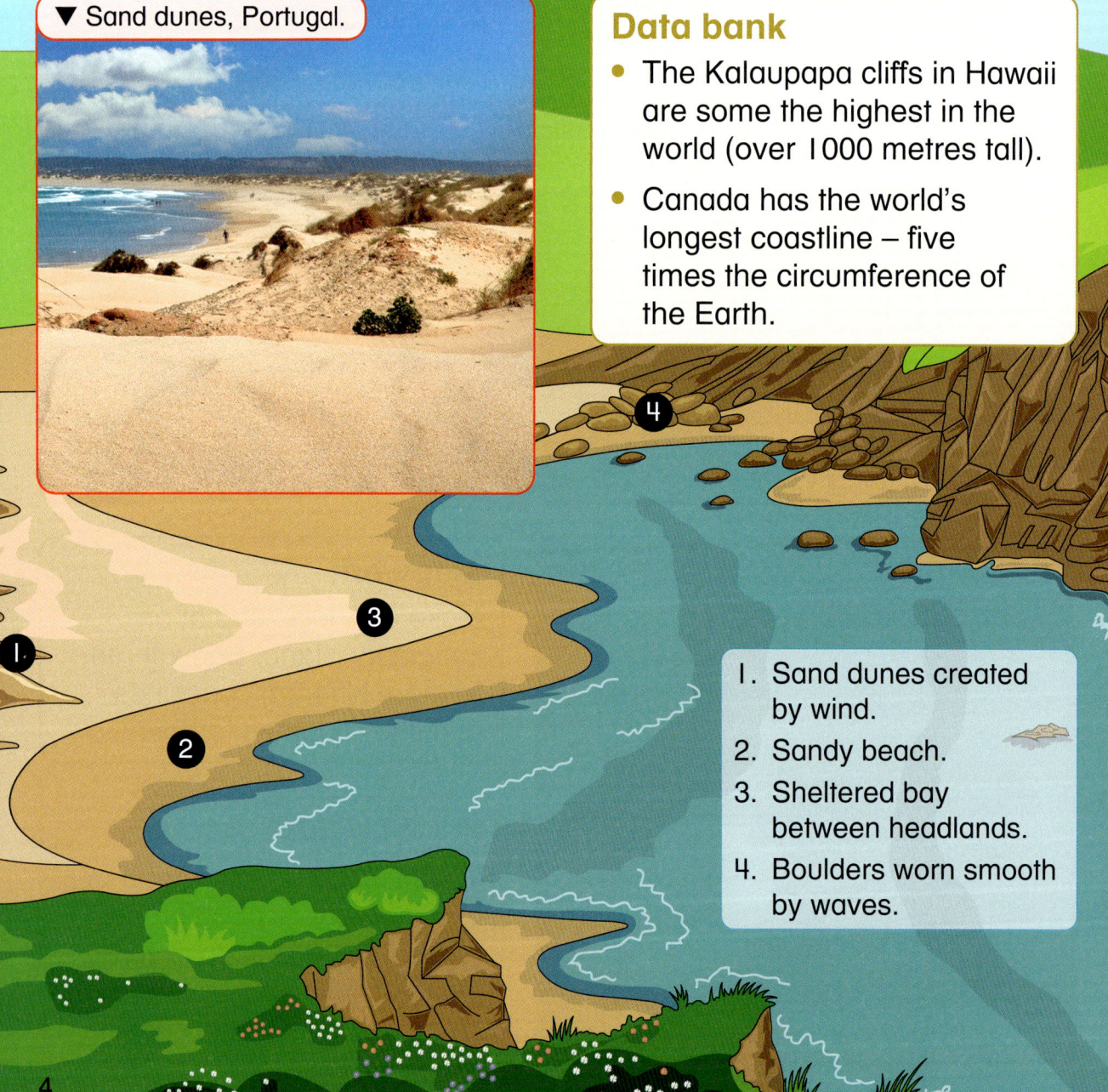

Data bank
- The Kalaupapa cliffs in Hawaii are some the highest in the world (over 1000 metres tall).
- Canada has the world's longest coastline – five times the circumference of the Earth.

1. Sand dunes created by wind.
2. Sandy beach.
3. Sheltered bay between headlands.
4. Boulders worn smooth by waves.

Unit 1 Coasts

Mapwork
Using an atlas, find three countries that have no coast at all.

Investigation
Make drawings with short notes of three seashore features in your geography notebook.

Discussion
- What shapes the coast?
- What different coastal features can you see in the drawing?
- How do you think the coastline in the picture might change next?

▼ Cook Strait, New Zealand.

5. Headland with cliffs juts out into sea.
6. Caves at the bottom of the cliff.
7. Cliff fall.
8. Rock stack.

Unit 1 Coasts

Lesson 3: Exploring the coast

How do people look after the coast?

Sidmouth

Sidmouth is a popular seaside town on the Jurassic coast in southern England. People go there because of the beautiful coastline and the sheltered, sunny beach. The map shows how the beach and the cliffs are looked after so visitors can enjoy them.

Investigation
Research online to find out about the Jurassic coast.

Key words
beach
cave
habitat
landslide
tide
zone

▲ The beach at Sidmouth, Devon.

Mapwork
Make up a key for the map of Sidmouth showing what the colours stand for.

Unit 1   Coasts

What can you find on the seashore?

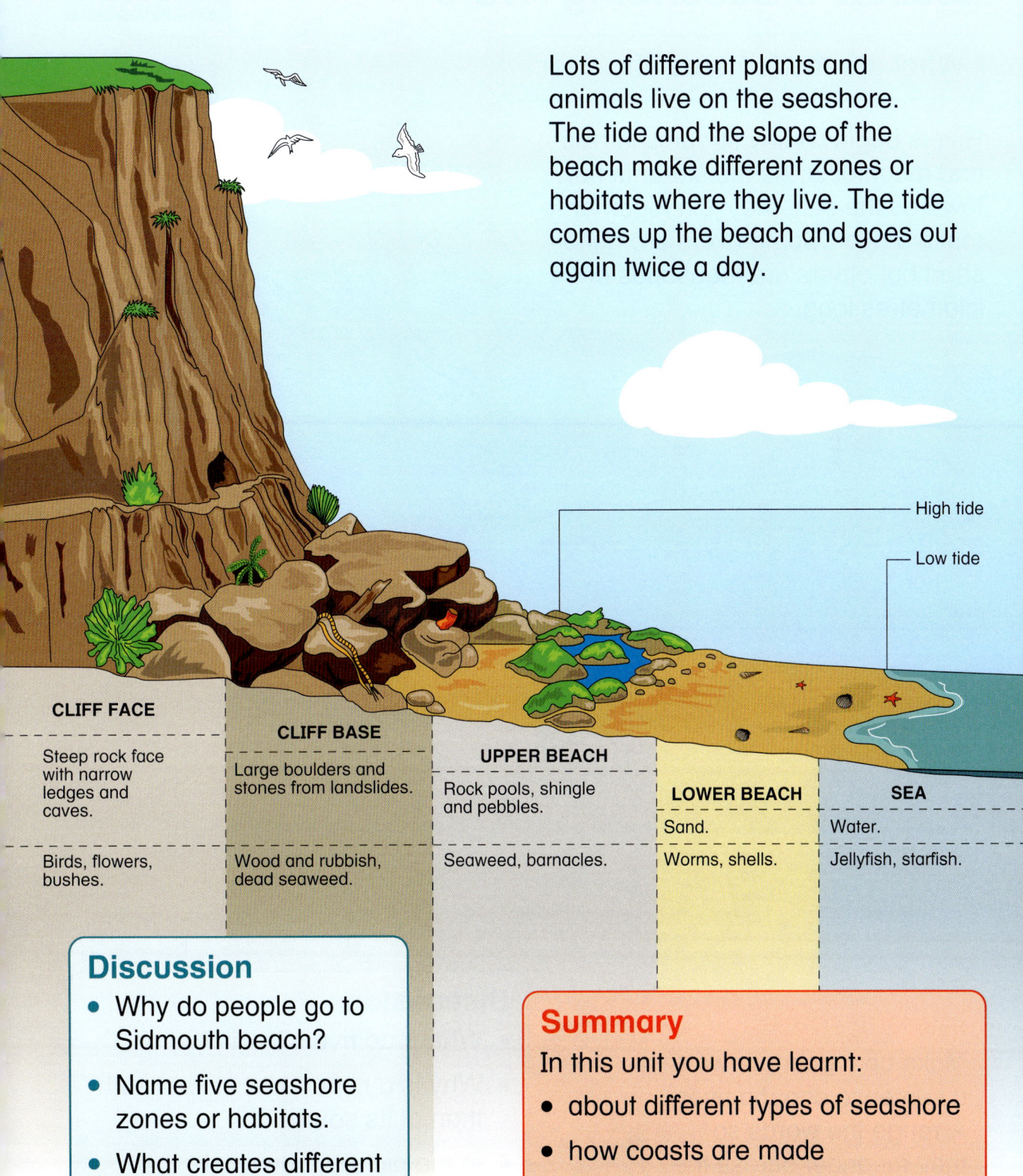

Lots of different plants and animals live on the seashore. The tide and the slope of the beach make different zones or habitats where they live. The tide comes up the beach and goes out again twice a day.

- High tide
- Low tide

CLIFF FACE
Steep rock face with narrow ledges and caves.

Birds, flowers, bushes.

CLIFF BASE
Large boulders and stones from landslides.

Wood and rubbish, dead seaweed.

UPPER BEACH
Rock pools, shingle and pebbles.

Seaweed, barnacles.

LOWER BEACH
Sand.

Worms, shells.

SEA
Water.

Jellyfish, starfish.

Discussion
- Why do people go to Sidmouth beach?
- Name five seashore zones or habitats.
- What creates different seashore habitats?

Summary
In this unit you have learnt:
- about different types of seashore
- how coasts are made
- about seashore habitats.

Unit 2 Rivers

Lesson 1: Describing rivers

What are the features of a river?

Key words
channel
estuary
gorge
meander
mouth
pool

Rivers begin as tiny streams in hills and mountains. As the streams flow towards the sea, they join together to make a river. Some rivers are quite short but others are thousands of kilometres long.

Source or place where the stream begins

Pool where water collects

Other streams called tributaries join the main stream

The river cuts a deeper channel

Investigation
Make up a poem or write a sentence about a river. Arrange the words so they meander across the page like a flowing river.

Discussion
- Where do rivers begin?
- Why is a river wider at its mouth than at its source?
- In the picture, how might the people in the town use the river?

Unit 2 Rivers

The salmon

The salmon was born in a pool high up in the hills near where the river begins. Over the weeks it slowly grew bigger and stronger. One day it swam into the tiny stream which flowed out of the pool. It did not know why, but it felt it needed to reach the sea.

Soon the salmon felt the stream getting deeper as water drained into it from the land. The stream was turning into a river. It began to flow faster. The salmon was bounced over rocks by the rushing water and it found it difficult to breathe in the tossing foam.

When the river reached flatter land it started to meander through countryside. The salmon swam on. Sometimes it swam with the strong current, sometimes it rested in the shallow water by reedy banks.

The salmon felt the river was changing. The river seemed to be deeper at some times of the day than others. The water even tasted different. It was salty now.

The salmon swam faster. It had reached the estuary with its banks of mud and small islands. Waves began to break on the surface of the water. The salmon's river journey was over. It had reached the sea.

Mapwork

Using a map or atlas, find a river in your area. Draw a diagram map to show its key features.

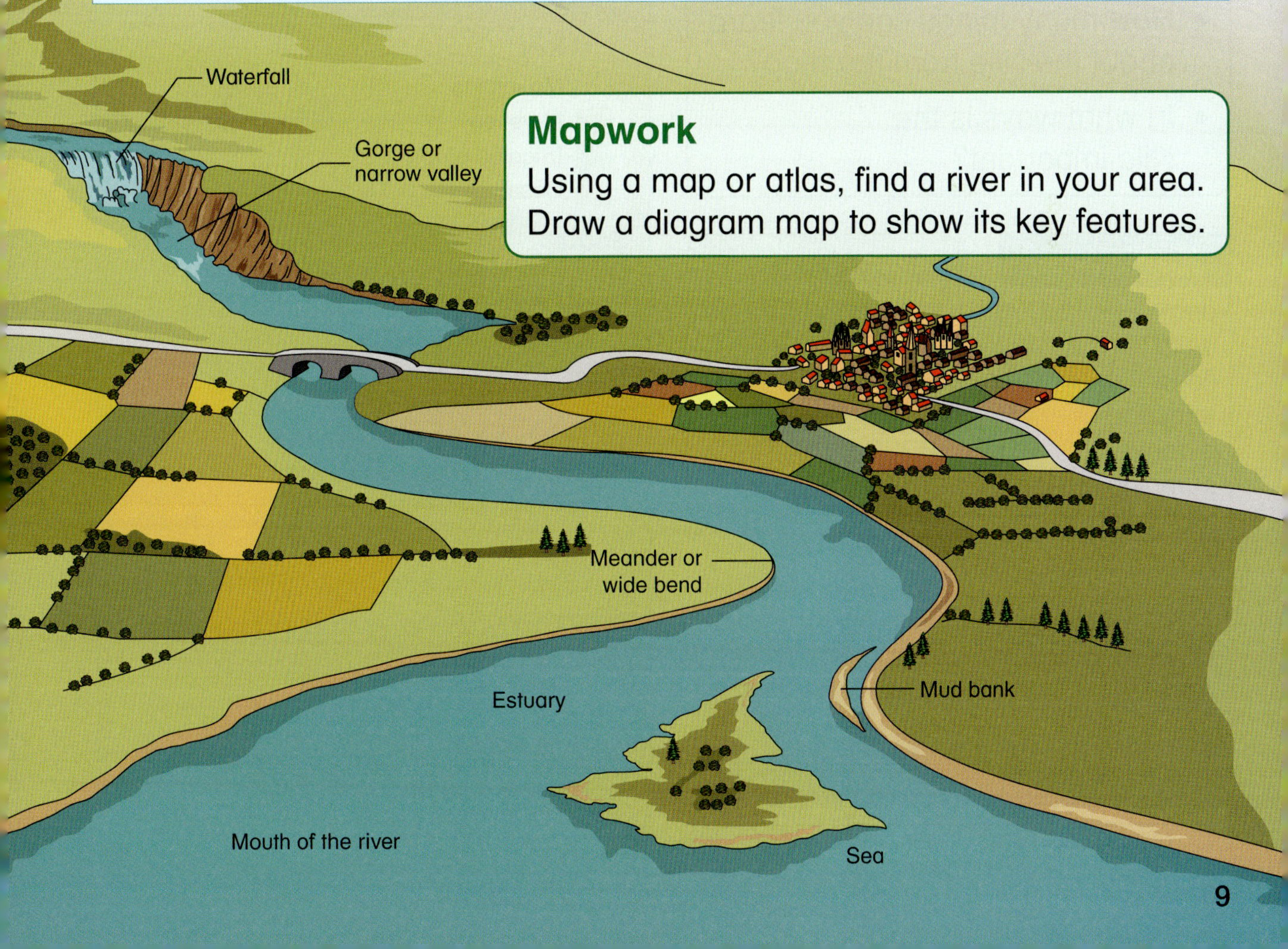

Unit 2 Rivers

Lesson 2: Rivers matter

How do people use rivers?

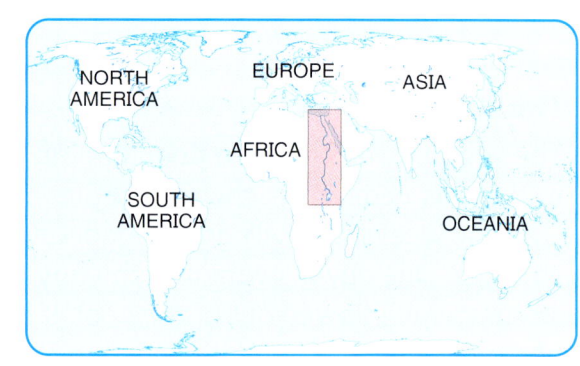

The River Nile is the longest river in the world. It rises in the mountains of Africa and flows north for nearly 7000 kilometres.

The Nile starts as two rivers. The Blue Nile rises in the mountains of East Africa. The White Nile comes from Lake Victoria. The two rivers join together before flowing across the Sahara Desert to the sea.

Discussion
- Where does the Nile begin?
- How many things can you learn about the Nile from the map?
- In what ways is the Nile important?

Key words
dam
delta
flood
irrigation
pyramids
pumping station

Mapwork
Using an atlas, name the continent for each river in the bar chart.

World's longest rivers

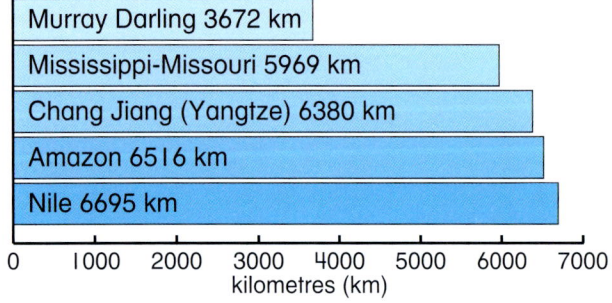

Murray Darling 3672 km
Mississippi-Missouri 5969 km
Chang Jiang (Yangtze) 6380 km
Amazon 6516 km
Nile 6695 km

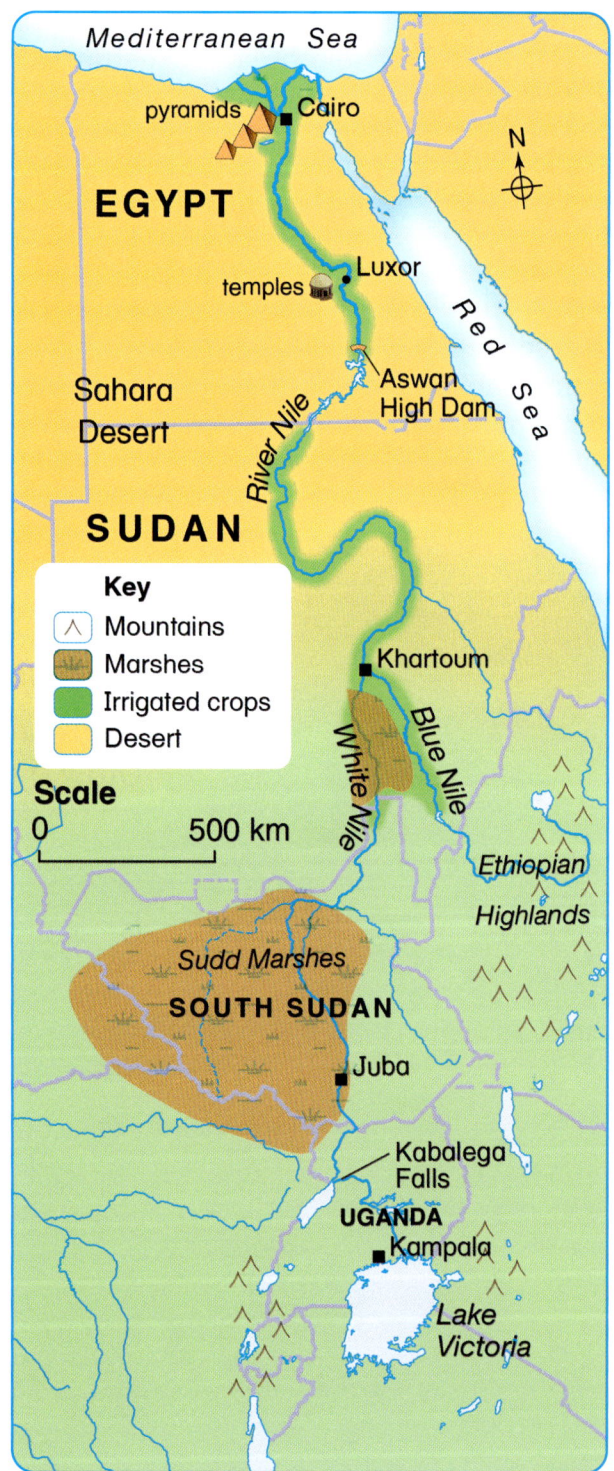

Unit 2 Rivers

The River Nile brings water to the desert. For thousands of years, farmers in Egypt and Sudan have used river water on their crops. This is called irrigation.

Many tourists visit the Nile valley. Some travel to the old temples and pyramids by boat. Higher up the river huge dams help to control floods.

Data bank
- As the Nile flows across the desert, a lot of water either sinks into the ground or evaporates.
- The Nile splits into small rivers when it reaches the sea, making a triangle of marshes called a delta.

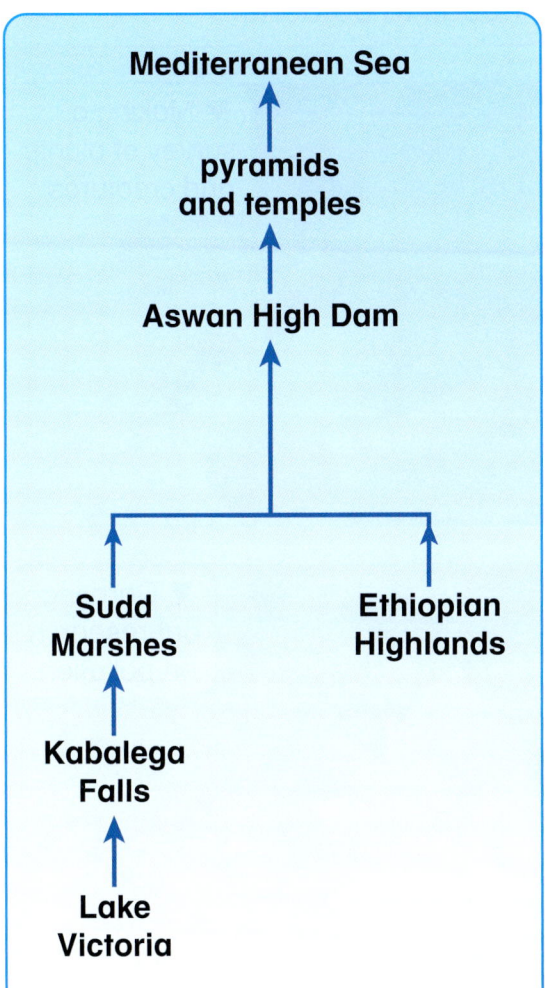

▼ Floating pumping stations take water from the river for the crops.

▼ Traditional sailing boats called feluccas still travel along the Nile.

▼ The Aswan High Dam uses the power of the river water to make electricity.

Investigation
Make a picture map for tourists showing the places they would see on a journey up the River Nile.

Unit 2 Rivers

Lesson 3: Managing rivers

How do people care for rivers?

Valerie Hay and her team work for the Environment Agency in northeast England. They help to keep the River Tees clean and healthy. These are some of the jobs that the team does.

Investigation

Write a short report about the work done by Valerie and her team. Say why each task needs to be done.

▼ Taking river water samples.

▼ Making a survey of plants and creatures.

▼ Making sure farm waste does not seep into rivers.

▼ Carrying out research on pollution.

Key words

bank
levee
sluice gates

Discussion

- What do Valerie Hay and her team do?
- What things can stop a river being clean and healthy?
- In what ways are rivers useful?

Unit 2 Rivers

Investigating the Sumida River, Tokyo

Children visiting Tokyo wanted to find out about the Sumida River. They went on a walk along the river bank to find out how people were using the water. They also made notes about the plants and creatures they saw. Back at school they continued their research. These are some of the things they discovered.

▲ Gulls, herons and ducks enjoy living by the water.

▼ Many boats use the river, including garbage boats.

▲ Sluice gates control the flow of water.

▲ Bridges help people to cross the river.

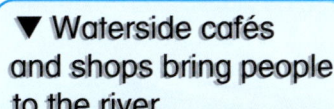

▼ Waterside cafés and shops bring people to the river.

▲ There is a walkway on top of the flood bank or levees.

Mapwork
See if you can find a map of a river walk in your area.

Summary
In this unit you have learnt:
- about the features of a river
- how people use rivers
- how to study a river.

Unit 3 — Weather patterns

Lesson 1: Extreme weather

How does the weather affect us?

The weather has a big effect on our lives. It affects the way we dress, the houses we live in and the things we do.

People are used to dealing with the weather. However, extreme weather sometimes causes serious problems.

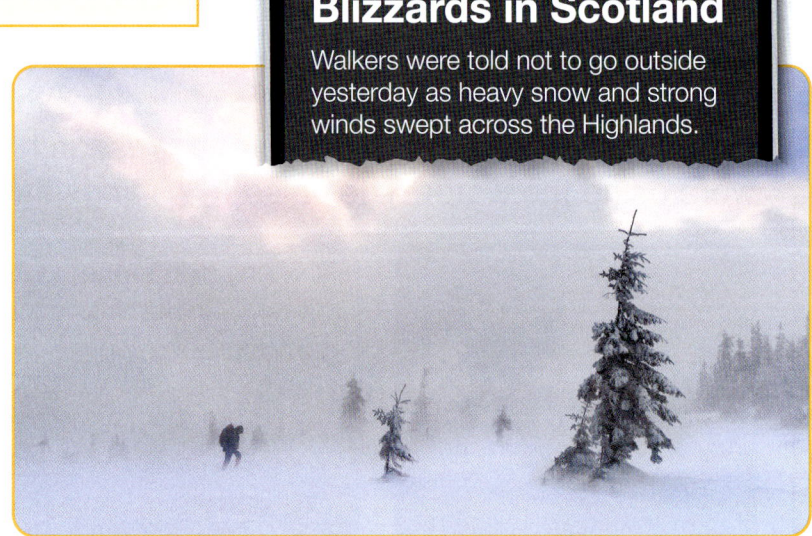

Blizzards in Scotland
Walkers were told not to go outside yesterday as heavy snow and strong winds swept across the Highlands.

Hurricanes in the Caribbean
Powerful winds and heavy rain are causing floods, destroying buildings and uprooting trees as hurricanes hit the Caribbean this autumn.

Discussion
- How does the weather affect us?
- What are the main problems shown in the photographs?
- How are people in your country affected by extreme weather?

Key words

blizzard	flood
bushfire	hurricane
drought	monsoon

Unit 3 Weather patterns

Data bank
- In 2021 a 'heat dome' formed over British Colombia, Canada, bringing temperatures of nearly 50 °C.
- Heavy monsoon rains in 2022 caused the River Indus in Pakistan to flood, affecting 33 million people.
- In 2023 Greece suffered one of the worst wildfires ever recorded in Europe.

Drought in Africa
Many countries on the edge of the Sahara Desert continue to be affected by years of drought.

Floods hit Germany yesterday
Extreme storms have caused rivers to reach their highest level in years in many parts of Germany.

Bushfires in Australia
Months without rain and strong winds are making it impossible for firefighters to stop bushfires spreading in southeast Australia.

Investigation
How might snow, floods, gales or drought affect (a) a tree (b) a house? Draw some 'before' and 'after' pictures.

Climate change
Scientists believe that climate change will make the weather more extreme.

Unit 3   **Weather patterns**

Lesson 2: Weather forecasts

Who uses the weather forecast?

Most of us are interested in what the weather is going to be like. Some forecasts give information for the next few hours. Other forecasts give a more general idea of what could happen over the next week. People can choose the type of forecast they need.

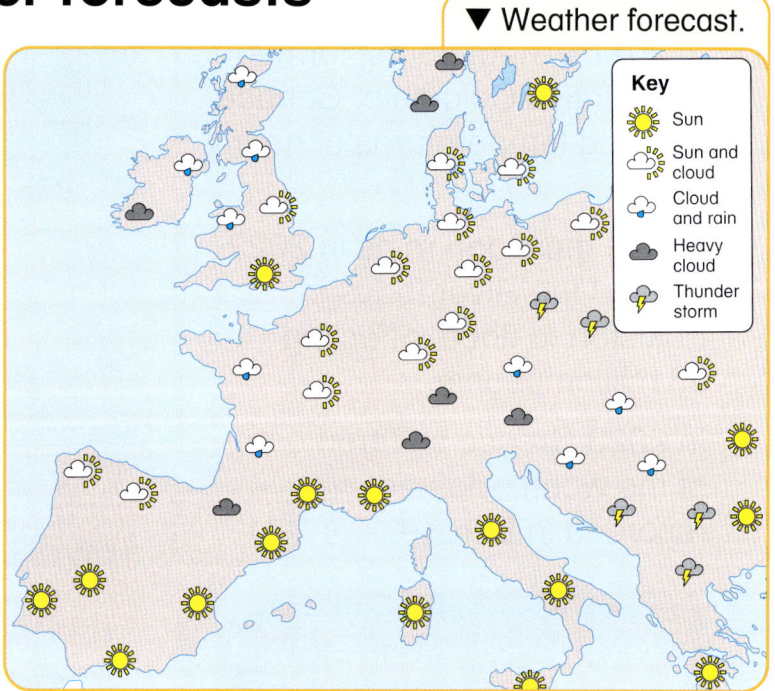

▼ Weather forecast.

Key: Sun, Sun and cloud, Cloud and rain, Heavy cloud, Thunder storm

> **Discussion**
> - Why is it useful to have weather forecasts?
> - Why do we need so many ways of gathering information?
> - How might the job you want to do in the future be affected by the weather?

▲ **Farmer**
"I wonder if the weather will be good enough to harvest the crops this week?"

▲ **Crane driver**
"I need to know how strong the wind will be this afternoon."

▲ **Family**
"If it's sunny tomorrow, can we go for a picnic?"

Unit 3 | Weather patterns

How are weather forecasts made?

Key words
forecast
temperature

Balloons
Weather balloons record information as they rise through the air. They send back messages by radio.

Satellites
Weather satellites take photographs of the clouds covering the Earth.

Aircraft
Aircraft carry instruments that give information on the winds along their flight paths.

Land stations
Each day people measure the temperature, the direction of the wind and the amount of rain which has fallen.

Weather centre
In the UK, weather data is collected by the Met Office and put into supercomputers

Ships
Ships record what the weather is like at sea.

Data bank
- The world's highest weather station is on Mount Everest at 8000 metres.
- Some weather satellites are stationed at 35 000 km above the ground.

Mapwork
Research today's weather map for your location. Write a few sentences saying what it tells you.

Investigation
Make up a weather forecast for the farmer, crane driver or family. How will it affect what they do next?

Unit 3 Weather patterns

Lesson 3: Recording the weather

How can we record the weather?

Key words
picture scale
rain gauge
symbol
thermometer
weather vane

At Hillside School the children decided to keep a record of the weather. They made their own weather recording instruments. They then kept records every day for two weeks.

▲ The children made a rain gauge, weather vane and other equipment to record the weather.

Discussion
- What are the three main parts of a weather record?
- Which of the three picture scales below would work best without words to go with it?
- How could you describe the weather today in six words or less?

Wind
There are clues which tell you about the strength of the wind.

Temperature
A thermometer measures the temperature in degrees.

Cloud
Different symbols are used to show sun, cloud, rain and snow.

Unit 3 Weather patterns

The children recorded the information they had collected and wrote a report about weather in their area.

▼ These fluffy cumulus clouds are often a sign of good weather.

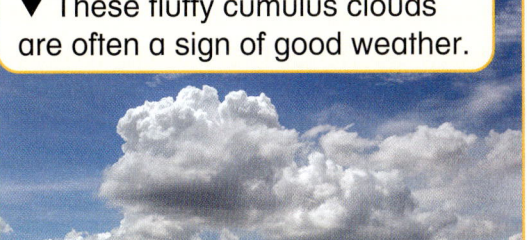

▼ The weekly weather diary.

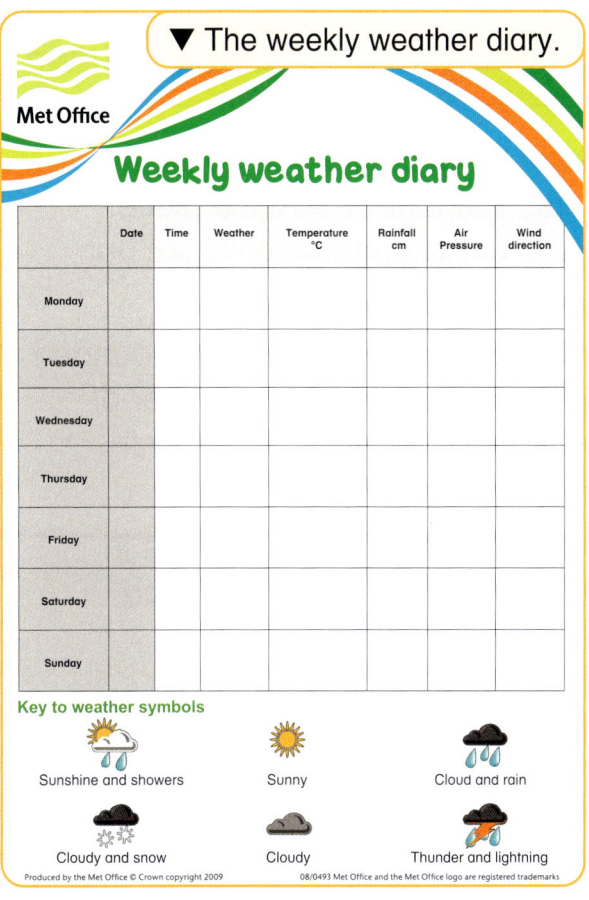

Mapwork

Select a city in any part of the world which begins with the same letter as your name. Using data from the internet, draw graphs and write a report about the weather there over a week or two weeks. Put your work around a world map in a class display.

Data bank

- The record for the hottest day is 57 °C, recorded in Death Valley, US in 1913.
- The wettest ever day was in the island of Reunion in the Indian Ocean when 1800 mm of rain fell in 24 hours.
- On average it rains one day in three in England.

Investigation

Record the weather at your school for two weeks. Show the results on a chart and write a short report. Did the weather affect the things you could do?

Summary

In this unit you have learnt:
- why we need weather forecasts
- how weather forecasts are made
- how you can record the weather.

Unit 4 Towns

Lesson 1: Understanding towns

What are the features of a town?

Towns are places where lots of people live together in one place. Towns are bigger than villages but not as large as cities.

East Kilbride in Scotland was founded in 1947. It was one of a number of new towns that were built after the Second World War.

People who lived in old houses in Glasgow and other cities moved there to make a new life.

When people plan a town they have to decide what to put in it.

These are some of the things which are needed:

- homes for families and single people
- places to work and shop
- schools and colleges
- doctors, dentists, hospitals
- places for sport and entertainment
- ways of moving around.

Investigation

Working from the photograph and the map on page 21 list all the places which are to do with (a) transport (b) health and education.

Discussion

- What makes East Kilbride a town not a village?
- What are the three most important parts of a town?
- What new things would you want to add to East Kilbride?

Mapwork

Using the map, name each of the places shown in the photograph. Then write down the grid square code.

Data bank

- East Kilbride is one of 22 'new towns' built in the UK after the Second World War.
- New towns were built to give people from cities like London a better life.
- Around two million people now live in new towns.

Unit 4 Towns

▲ Map of East Kilbride.

Key words

grid square
education
health
transport

Unit 4 Towns

Lesson 2: The origin of towns

How did towns begin?

Key words	
castle	mine
factory	port
market	resort

Many towns started off as small places which then grew larger. Some towns, like Jericho and Damascus in the Middle East are extremely old. Others have developed much more recently.

Towns are always changing. As they expand, new houses, shops and schools are built around the edge. Sometimes the centre is rebuilt. Towns which go on growing eventually turn into cities. Today more people are living in towns and cities than ever before.

▲ **Market towns**
Many towns began as places where people came to trade their crops and animals. Djenné in Mali is an ancient trading centre on a route across the Sahara Desert.

Data bank
- Damascus in Syria was founded around 10 000 ago.
- The first towns in the UK were built by the Romans.
- Towns often have between 1000 and 100 000 people in them.

Discussion
- How did many towns begin?
- What are the five different types of town shown in the photographs?
- Which type of town would you most like to visit and why?

▲ **Crossing points**
Some towns grew up around bridges or places where routes come together. Istanbul in Turkey, is on the crossing point between Europe and Asia.

Unit 4 Towns

▶ Factory towns
Factory towns were created for people who worked in factories or mines. Lowell, in the US was built by the owner of textile factories.

◀ Seaside resorts
Resorts are places where people go to stay to be at the seaside. The sunshine and beach attract tourists to Cancun, Mexico.

▶ Ports
Towns are often built around harbours where ships can load and unload their cargoes safely. Dover is a port town in the UK.

Mapwork
Using a road atlas or online map, look at a street plan of your nearest town. Write down any clues which tell you what type of town it is.

Investigation
Make a list of any towns in your area which have (a) a castle (b) a harbour or (c) a bridge over an important river.

Unit 4 Towns

Lesson 3: Town life

How does a town work?

▲ Children working in the street.

In the past, people who lived in the country grew their own food, fetched water from a well and gathered wood from the forest. They were able to look after their own needs.

Then more and more people started moving to the towns to find work. Water, food and other services had to be provided for them.

Discussion
- How did people look after their own needs in the past?
- What services keep a modern town working?
- What would happen if one of the services stopped working? Think about each one in turn.

Key words
power
services
transport
waste disposal

Things which keep a modern town working

Unit 4 Towns

The children at the International school went for a walk near their school. They looked for clues of different services.

These are some of the things they found. Think about how each one helps people in their daily lives.

▲ Litter bin

▲ Grate for water meter

▼ Street sign

▼ Bus stop

▼ Post box

▲ Bench

Investigation
Make a similar survey in streets near your school. Take photographs of the different things you discover. Say how each one helps to keep the town working.

▲ Road barriers

Mapwork
Where would be the best place for a new litter bin in your area? Show your ideas on a local street map and say why you have chosen that place.

Summary
In this unit you have learnt:
- about the buildings and places in a town
- how towns change over time
- how towns provide the things people need.

Unit 5 Food and shops

Lesson 1: Farms and food

Where does food in the UK come from?

Key words	
crops	food miles
dairy	soil
landscape	

Farmers in the UK produce a great variety of food. Some parts of the country are good for growing crops, fruit and vegetables. Other parts are better for keeping cows and sheep.

When farmers decide how to use their land, they have to think about four main things.

Weather — Is there the right mix of rain and sun?

Landscape — Is the land too hilly to grow crops?

Soil — Is the soil right for the plants?

Market — Will people pay a good price for the produce?

▼ **Sheep farms**
In the hilly areas of Wales and Scotland, many farmers keep sheep.

▼ **Dairy farms**
In western and central England and Ireland, grass grows well and farmers keep cows.

▲ **Crop farms**
In eastern England there is more sunshine and less rain so farmers grow wheat and other crops.

Key
- Cows
- Sheep
- Crops

26

Unit 5 Food and shops

Where does food come from?

The food in UK shops comes from farms all over the world.

Data bank
- Three-quarters of the land in the UK is used for farming.
- A cow produces about 25 litres of milk a day.
- There are around 22 million sheep in the UK and over 50 breeds.

Investigation
Looking at the shop picture, make a list of food that has come from (a) Great Britain (b) other countries.

Discussion
- What are the three main types of farm?
- Which of these farms is in the countryside in your area?
- What are three benefits of buying local food?

Mapwork
Work out how far some of the food sold in the shop will have travelled to the UK or to where you live. Use an atlas to find the 'food miles' between capital cities.

Unit 5 Food and shops

Lesson 2: From farm to supermarket

> How does food get to a supermarket?

Jane Ford works for a company which imports fruit into the UK. Her company buys fruit from different countries and sells it to supermarkets. Each day three large lorries arrive from the docks on the River Thames. The fruit is unloaded as it arrives.

Jane has been doing her job for many years. *"In the past you could only get fruit and vegetables when they were in season,"* she explains. *"Now you can buy them all the time. When the harvest stops in one country it starts in another."*

Key words
- import
- Fairtrade
- plantation
- season
- Caribbean

Discussion
- What does Jane's company do?
- What are the stages of bringing bananas to UK supermarkets?
- What is good and bad about being able to eat fruit and vegetables out of their season?

Data bank
- On average each person in the UK eats 100 bananas a year.
- One in three bananas sold in the UK is 'Fairtrade'.

Importing bananas

These are the stages in importing bananas from the Caribbean. Most of the bananas sold in the UK come from this region.

Growing

The bananas are grown on a plantation.

Packing

They are put into boxes in a pack house.

Unit 5   Food and shops

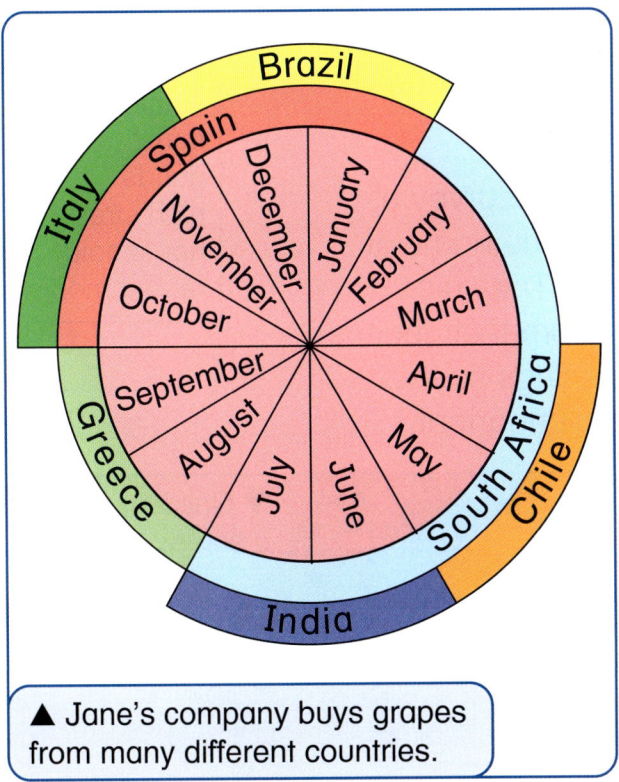

▲ Jane's company buys grapes from many different countries.

▼ The grapes are checked for quality.

Investigation
Make a zigzag book of the different stages of getting bananas from the Caribbean to the UK or to your country.

Mapwork
Using an atlas work out the route a lorry with oranges might take from Turkey to the UK or to your country. What countries and cities would it pass through?

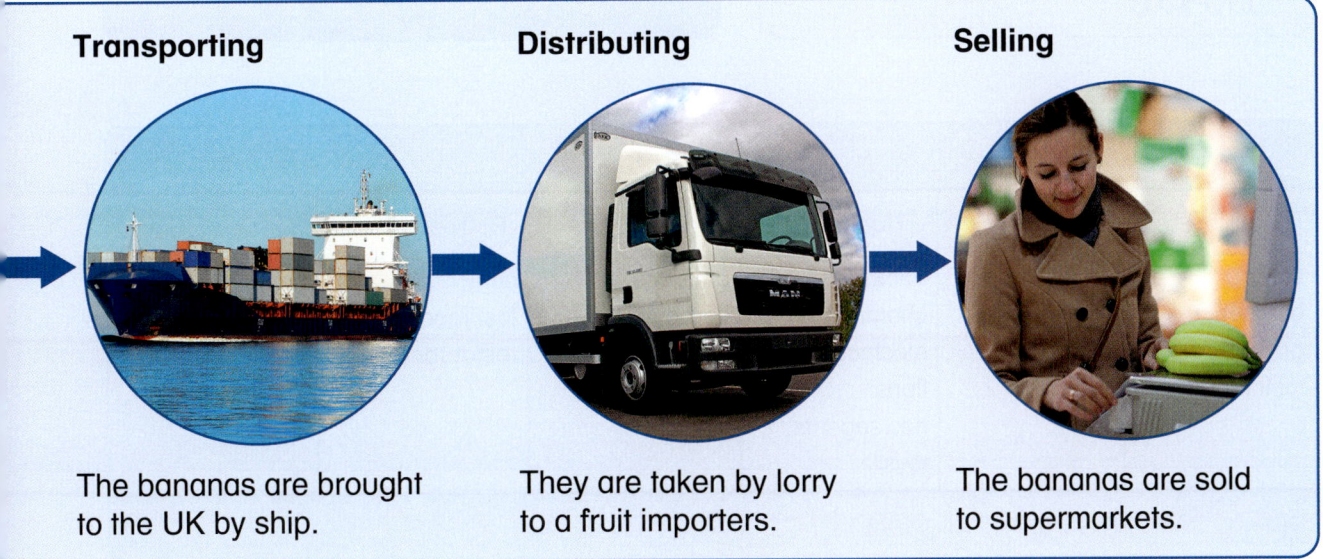

Transporting
The bananas are brought to the UK by ship.

Distributing
They are taken by lorry to a fruit importers.

Selling
The bananas are sold to supermarkets.

Unit 5 Food and shops

Lesson 3: Local shops

Investigating local shops

At Benwell in Newcastle the children made a survey of shops in their local high street. They wrote the name of each shop and the main things that it sold. They also sorted the shops into groups using a list from their teacher. Some of the shops sold a lot of different things so the children had to decide what was the most important.

Key words
antiques
estate agent
florist
goods
high street
household
refreshments

Investigation
Make a similar survey of shops near your school.

Data bank
- Around a quarter of the goods sold in the UK are bought online.
- UK high streets are changing as some shops close and others change hands.
- The first supermarket in the UK opened in London in 1951.

▼ Children doing a survey.

Type of shop					
Food	Clothes	Household	Furniture	Refreshments	Money/offices
baker	charity	books	antiques	café	bank
butcher	clothes	chemist	carpets	fast food	estate agents
greengrocer	shoes	electrical	furniture	restaurant	post office
mini market	sports	florist			travel agents
		newsagents			
		toys			

Others: hotel, hairdressers, music shop

Unit 5 Food and shops

When they returned to school the children were given large scale maps of the high street. They coloured the key and the shop outlines using a different colour for each type. They also added up the totals to find which type of shop was most common and made a bar chart to show their results.

Shops in Benwell

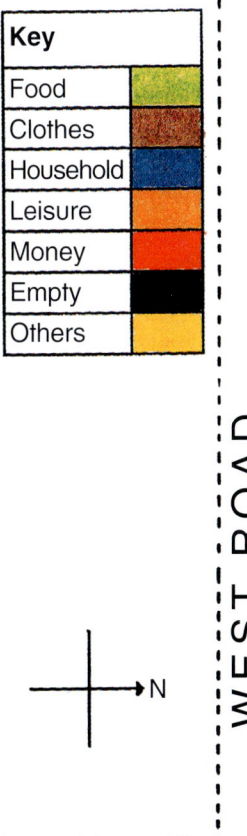

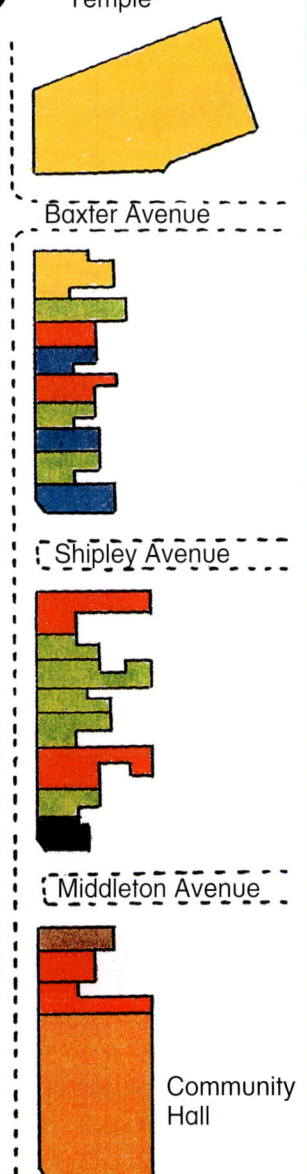

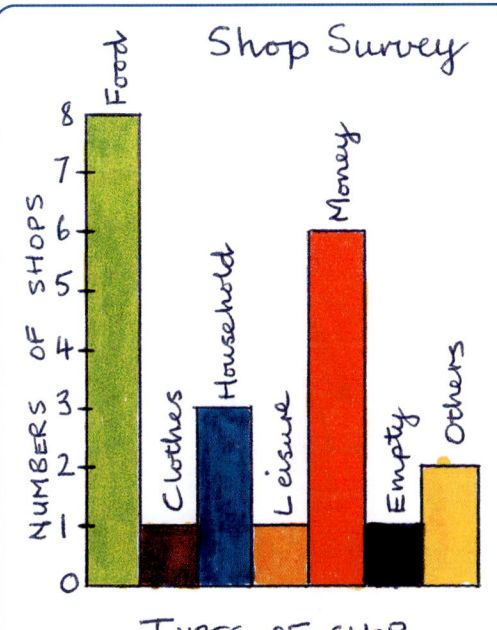

We found food, household and money shops were the most common.

Discussion
- Why are high streets changing?
- Which type of shop do you visit most?
- Which shops do you think are most important?

Summary
In this unit you have learnt:
- why there are different types of farm
- that supermarkets sell food from all over the world
- how to study local shops.

Unit 6 — Caring for towns

Lesson 1: Old and new buildings

> What happens to old buildings?

Most of us think that the house where we live and streets where we grow up are special in some way. However buildings wear out. Most old buildings are knocked down and replaced by new ones. A few are saved because they are interesting or important for their history.

Key words
listed buildings
windmill

Discussion
- Why do you think the houses in the photograph are being pulled down?
- Which are the oldest buildings in your area?
- Why do you think the buildings in the photographs on page 33 were saved?

Data bank
- Buildings which are very old or special in some way are protected by law. They are called listed buildings.
- There are around 370 000 listed buildings in England.

Unit 6 Caring for towns

◀ In India this old palace has been turned into a luxury hotel.

▶ This old coal mine in Poland is now used for meetings and conferences.

▼ This windmill in Greece is now a home.

Mapwork
Draw plans to show how you could turn a windmill with three floors into a house.

Investigation
Find out about buildings in your area which have been protected. Arrange to visit them so you can see why they are so special.

Unit 6 Caring for towns

Lesson 2: Making improvements

How can places be improved?

Key words	
bollard	pedestrianisation
chicane	town planner
park and ride	traffic calming

Duncan Brown works as a town planner in Glasgow. His job is to improve the streets and other places in the city. The Council has asked him to make more space for walking and cycling in the town centre. This will make it pleasanter and will reduce pollution from cars.

Duncan begins by making surveys and taking photographs. He then writes a report, draws plans of the changes and works out the cost. Next the Council asks local people what they think of Duncan's plans. This gives everyone a chance to improve the place where they live.

Cochrane Street

[Diagram with labels: trees, path, John Street, tower, path, benches, Ingram Street]

▲ Plans for improving Cochrane Street, Glasgow.

▲ People and cars in Cochrane Street.

Discussion
- What does Duncan Brown do?
- How do you think the changes made Cochrane Street pleasanter?
- What other improvements could have been made?

34

Unit 6 Caring for towns

Improvement schemes

Investigation
How does each improvement make the environment better? Write a short report to go with the drawings.

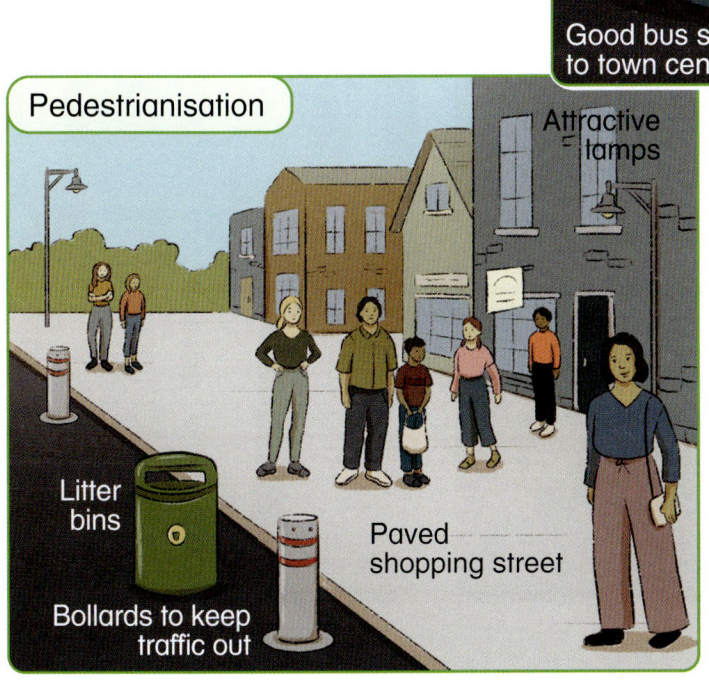

Mapwork
Find an area near your school which might be improved. Make plans and write notes about your ideas.

Unit 6   Caring for towns

Lesson 3: Comparing places

> Which place is best?

All over the UK there are competitions for the best kept towns and villages. When the judges visit each place they have a list of questions. They give marks to various features to decide which town will win the award.

Key words
- advertisement
- award
- graffiti
- gutter
- satellite dish
- street furniture

Buildings
Are the buildings in good repair?
- walls
- roofs
- gutters
- paintwork
- windows

Town A

Town B

Pollution
Is the place clean or dirty?
- litter
- graffiti
- oil stains
- fumes

Town A

Town B

Unit 6 Caring for towns

Town A   Town B

Street furniture

Are there enough things to help people?

- seats
- lights
- signs
- post boxes
- litter bins

Town A   Town B

Appearance

Is the place pleasant or unpleasant?

- overhead wires
- advertisements
- plants and trees
- satellite dishes
- noise

Discussion
- What are four ways of judging a building or street?
- Do you think Town A or Town B would win the best kept town award?
- If your school was in a competition, what would be its best features?

Climate change
New houses can be built with solar panels, water butts and walls that keep out the cold. Which of these do you think most matters?

Mapwork
Take photographs showing things which could be changed or improved in your area. Present your work as a class display around a large scale map of your local streets.

Summary
In this unit you have learnt:
- how old buildings can be given new uses
- how people look after towns and villages
- how to compare different environments.

Unit 7 — Northern Ireland

Lesson 1: Introducing Northern Ireland

What is Northern Ireland like?

Northern Ireland is the smallest country in the United Kingdom. It lies to the west of Scotland across the Irish Sea.

Key words
- Belfast
- Lough Neagh
- River Bann
- Slieve Donard
- Sperrin Mountains

▼ The Giant's Causeway in Northern Ireland is made of thousands of hexagonal rocks.

Key
- Over 500 metres
- 200–500 metres
- 0–200 metres

Scale 0 20 40 60 km

Discussion
- How would you tell someone where Northern Ireland is?
- What is distinctive about the weather, settlement and work in Northern Ireland?
- Where would you like to go if you went to Northern Ireland?

▲ When it was launched in Belfast in 1912, the *Titanic* was the largest ship in the world.

Unit 7 Northern Ireland

Rivers and landscape
The River Bann is the longest river. Gently rolling countryside covers most of Northern Ireland. In some places, there are rugged hills and mountains.

Weather
West winds from the Atlantic Ocean bring a lot of rain. In summer it is cool but there is not much snow in winter.

Settlement
Belfast, the capital, is the only large city in Northern Ireland. Londonderry (Derry), Bangor and Newry are smaller towns around the coast.
▼ Belfast City Hall.

Data bank
- Ireland is sometimes called the Emerald Isle because of its green landscape.
- Lough Neagh is the largest lake in the United Kingdom.

Investigation
Find some photographs of Northern Ireland. Decide on two you would use as a screensaver. Say why.

Mapwork
Make a map of ferry routes from Northern Ireland to the rest of the UK using an atlas or online.

Work
Ships have been made in Belfast for many years. There is also work in factories making aircraft, food and clothes. In the country, farmers produce meat, eggs and butter.

Transport
Road and rail routes spread out from Belfast. Motorways connect Belfast to Ballymena and Dungannon.
▼ Belfast docks.

Unit 7 Northern Ireland

Lesson 2: Living in Northern Ireland

What is it like to live in Northern Ireland?

Key words
barn
peat bog
pass

Sienna and Patrick O'Neill live at Ballyknock Farm in County Londonderry (or County Derry as many people call it). Their parents spend most of the time running the farm. Mrs O'Neill also has a part time job in a nearby clothes factory.

Sienna and Patrick go to school in Maghera. A bus comes to collect them at 8:00 a.m. There are about 350 children in the school. Most of them live in the countryside.

After school Sienna and Patrick help their parents on the farm. Sometimes they collect the cows for milking. In the summer they help to harvest the grass in the fields. They often have to wear anoraks because it rains a lot in Northern Ireland.

▲ The barns on the farm are used for storing hay.

▲ The bus stops near to the farm on the way home.

Unit 7 Northern Ireland

Ballyknock Farm is close to the main road from Belfast to Londonderry (Derry). From the farm you see the road winding across the Sperrin Mountains. Ballynahone Bog, a nature reserve, is a few kilometres away.

Sienna and Patrick have lived at Ballyknock Farm all their lives. Mr and Mrs O'Neill like the peace of the countryside but sometimes Sienna and Patrick wish more of their friends lived nearby.

Discussion
- What jobs do Mr and Mrs O'Neill do?
- How do Sienna and Patrick help on the farm?
- How is your life the same and different to Sienna and Patrick's?

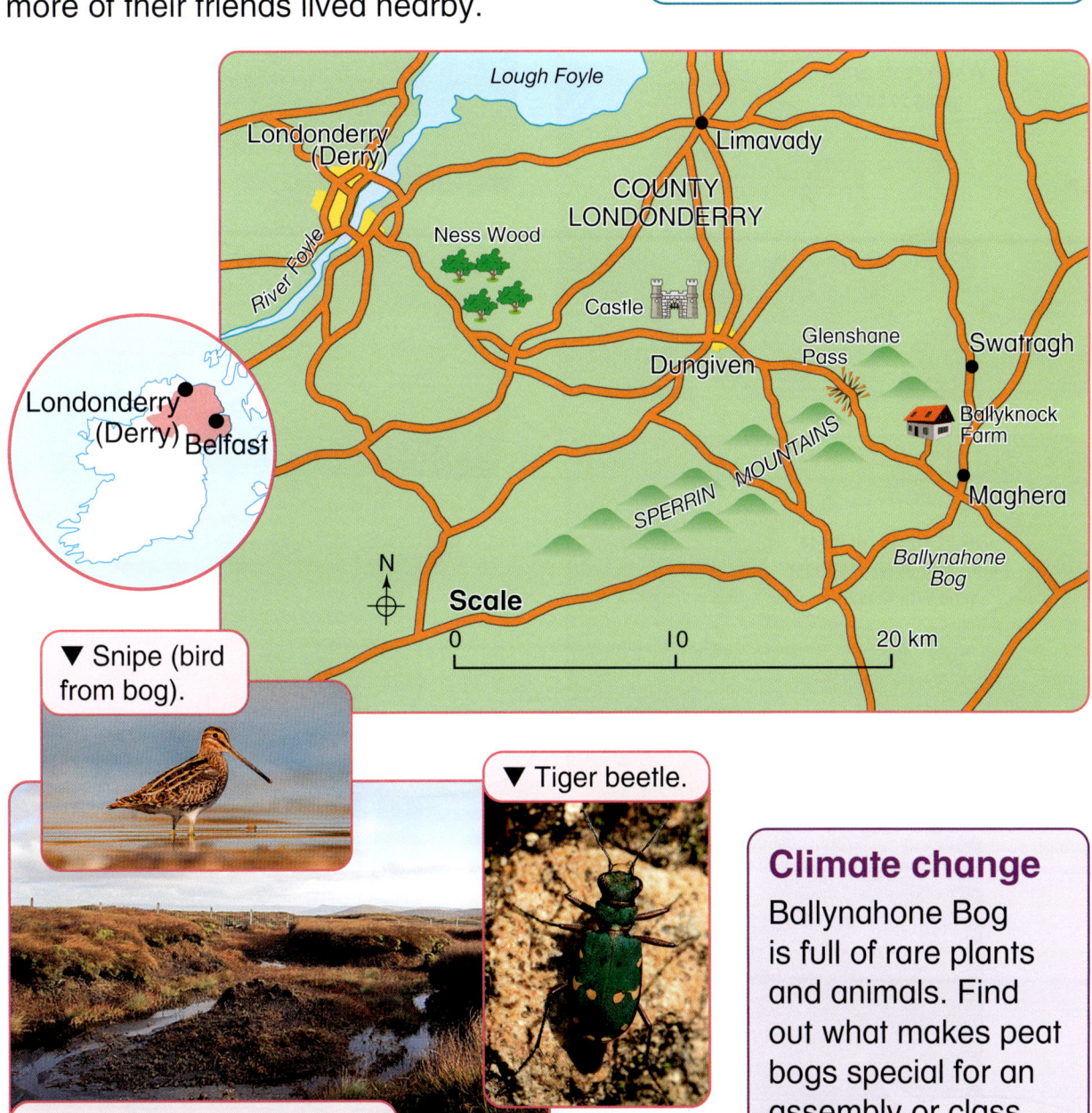

▼ Snipe (bird from bog).

▼ Tiger beetle.

▲ Rare plants and creatures live in the peat bogs.

Climate change
Ballynahone Bog is full of rare plants and animals. Find out what makes peat bogs special for an assembly or class presentation.

Unit 7 Northern Ireland

Lesson 3: A journey to Londonderry (Derry)

What could you see on a journey to Londonderry (Derry)?

On Saturdays the O'Neills sometimes go shopping in Londonderry, or Derry, as many people call it. The journey takes about 45 minutes. The drawings show some of the things they see on the way.

Unit 7 Northern Ireland

Key words
- goods
- Gaelic
- picture map
- port
- route

▶ The River Foyle and Londonderry (Derry).

Londonderry (Derry) is an important town with lots of factories, shops and a large hospital. Ships sail up the River Foyle to unload their goods at the port.

Sienna and Patrick like coming to Londonderry (Derry). Sometimes their father takes them to the cinema.

▲ Shipquay Street in the town centre.

Data bank
- Londonderry (Derry) is the second largest city in Northern Ireland.
- Many place names have Gaelic (Irish) roots. 'Bally' means farm or village and 'derry' means oak tree.
- About a sixth of Ireland is covered by bogs, creating a unique habitat.

Mapwork
Draw a picture map of the route from Ballyknock Farm to Londonderry (Derry). Add pictures or symbols of features along the way.

Summary
In this unit you have learnt:
- about the landscape and weather of Northern Ireland
- about the work and lives of different people
- why peat bogs are important to the environment.

Investigation
Design a brochure for tourists to encourage them to visit Londonderry (Derry).

Unit 8 Germany

Lesson 1: Introducing Germany

Germany

What is Germany like?

Germany is a country of north-central Europe. More people live in Germany than in any other country in Europe. For many years Germany was divided into two parts. It was united in 1990.

Rivers and landscape

The Rhine, Weser, Elbe, Danube and Oder are the main rivers. Northern Germany is mostly flat. There are hills and mountains in the south.

▼ The River Rhine goes through a deep gorge as it flows north from Frankfurt.

Settlement

Most people live in towns and cities. Berlin is the capital city.

▼ The Brandenburg Gate, Berlin.

Climate change

Climate change is bringing extreme weather all over the world. In 2021 entire villages were swept away in floods in north west Germany.

Work

Germany used to be famous for coal mining and for making steel. Today it makes machinery, cars, electrical goods and chemicals. There are also many banks and offices.

Investigation

Make up a quiz with five questions about Germany for other children in your class.

Unit 8 Germany

▼ The German flag.

Key words
Berlin
Hamburg
River Danube
River Rhine

Key
Mountains
Hills

Scale
0 100 200 300 km

Food
Baked pastries called pretzels are very popular in Germany.

Transport
Germany was the first country to build a network of motorways in the 1930s. It also has a high-speed train system.

Discussion
- What are the countries which border Germany?
- What are the differences between the weather in Germany and your country?
- What three questions would you like to ask about Germany?

Mapwork
Draw your own map of Germany showing two rivers, three cities and one mountain.

Unit 8 Germany

Lesson 2: The Ruhr: An industrial region

What is the Ruhr like?

Dominic is nine years old. He lives in Durham in the UK. His mother is German.

Dominic's relations live in Dinslaken in an area known as the Ruhr. This used to be one of the most important areas in the world for the iron, steel and coal industries.

The journey from Durham to Dinslaken takes Dominic and his family the whole day. First they drive to Harwich to take the ferry to the Hook of Holland. From there they take the motorway across the Netherlands to Germany.

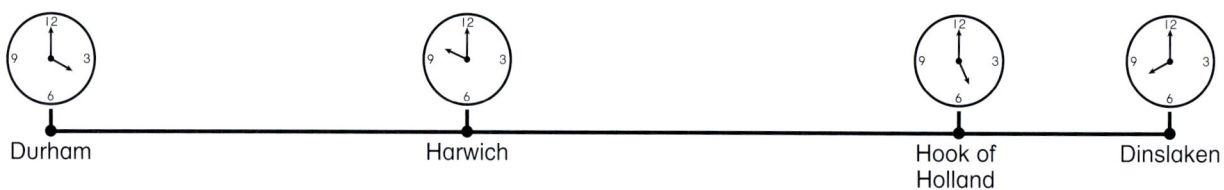

◀ A map of the Ruhr region.

46

Unit 8   Germany

Dinslaken: A town in the Ruhr

Just over 70 000 people live in Dinslaken. At the centre of the town there is a busy shopping street. During the summer it is warm enough to sit outside one of the cafés.

In the past many people used to work in the coal mine and iron works. Now there are new factories making metal goods using the latest technology. One hi-tech firm makes railway signals.

In the centre of Dinslaken there are large buildings such as the town hall, sports centre, college and banks. There are also markets for fresh fruit and vegetables.

▼ The Lohberg mine closed in 2006, after producing coal for 100 years.

Key words	
coal mine	railway signals
ferry	Ruhr
iron	steel
market	technology

Discussion
- How many hours does it take to get from Durham to Dinslaken?
- What was the Ruhr famous for?
- Why do you think the coal mines have closed?

Data bank
- There are 53 towns and cities in the Ruhr and around 5 million people.
- The Ruhr is named after the river which flows through it.
- In 1910 there were 400 000 miners in the Ruhr producing coal for much of Western Europe.

Mapwork
Working from the map, make a list of five towns in the Ruhr.

Investigation
Search the internet to find three interesting facts about the Ruhr.

Unit 8 Germany

Lesson 3: Living in Dinslaken

What is it like to visit Dinslaken?

When Dominic is in Dinslaken he always visits his cousins. Leon is ten years old and goes to a nearby school. Mia is six and goes to a nursery called a kindergarten.

Leon and Mia's father is a manager of the local supermarket. Their mother is a nurse. She works in the accident unit at the hospital.

When Dominic visits his cousins in the summer they often go to Rotbach Lake for a picnic. People go there to ride their bikes, play ball games or walk in the woods. Some people sail or fish. In the winter they go ice skating or watch an ice hockey match. Sometimes they go to the indoor swimming pool.

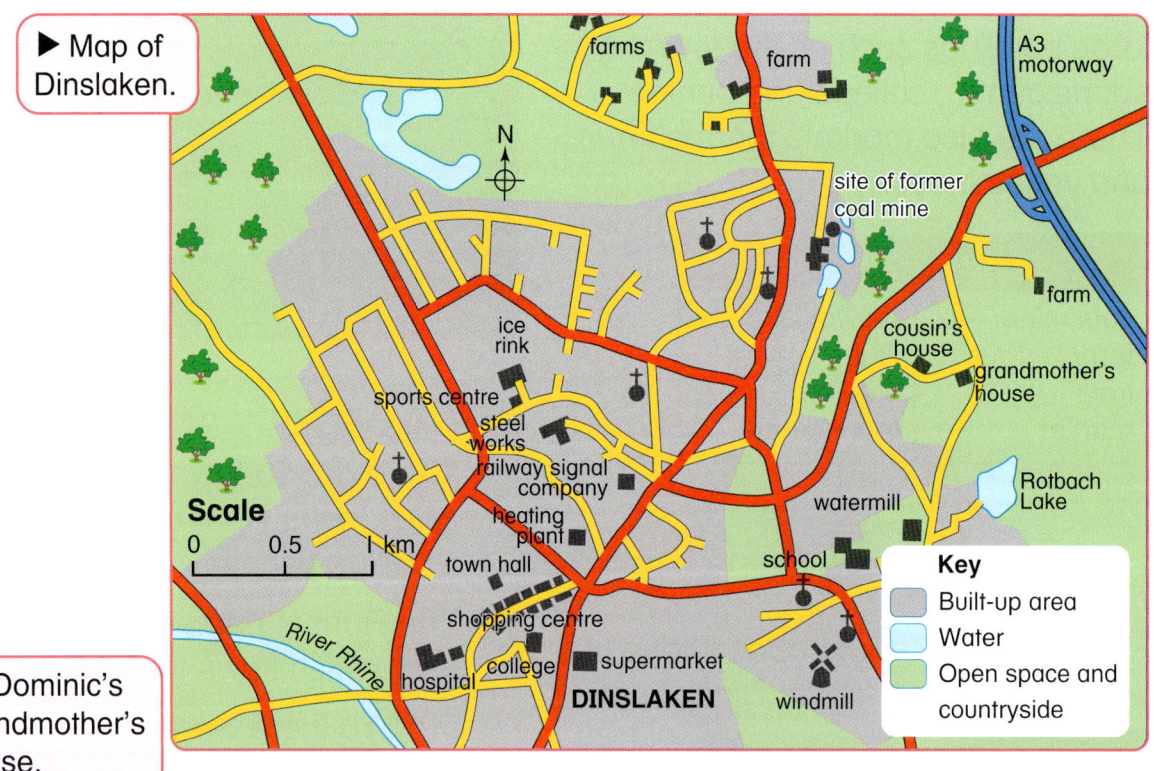

▶ Map of Dinslaken.

▼ Dominic's grandmother's house.

▶ Dinslaken was one of the first places in the Ruhr to have pedestrianised streets.

Unit 8 Germany

Changes and differences

Each time Dominic visits Dinslaken he notices changes. The coal mine has closed and some of the old factories have been pulled down. There is a new heating plant which burns the town's rubbish.

Sometimes Dominic looks out of the window and thinks of his home in England. The houses are not quite the same shape here, the signs are all written in German and there are no hedges to the fields. It is also colder in winter. Things are a little bit different in Germany.

Dinslaken in history

Dinslaken became an industrial town in the 19th century. The castle was built 700 years ago.

- **1850** Glue and steel factories set up
- **1856** Railway arrives
- **1894** Over 100 000 cattle brought to the market
- **1906** Coal mine opened to supply fuel to the ironworks
- **1945** Town rebuilt after being destroyed in the Second World War

Key words

heating plant	pedestrianised
kindergarten	watermill

◀ The watermill museum celebrates the history of the town.

Discussion
- Where is Dinslaken?
- What changes has Dominic noticed?
- What do you find most interesting about Dinslaken?

Investigation
Draw pictures of four events in Dinslaken's history and add dates to make a simple timeline.

Mapwork
Look at the buildings shown on the map of Dinslaken. Make a list of some of the jobs people might do there.

Summary
In this unit you have learnt:
- about the landscape and weather in Germany
- about a region in Germany
- about living in Germany.

Unit 9 — North America

Lesson 1: Introducing North America

What is North America like?

The landscape of North America is very varied. There are rainforests in Mexico and the countries around the Caribbean Sea. The United States (US) has grasslands, deserts and mountains. In Canada there are many lakes and great expanses of coniferous forest.

The US dominates North America. It has many industries and resources and is one of the richest countries in the world. Many of the people who live in the US have family links with Europe and other continents. There are also small numbers of American Indians who have lived in North America for thousands of years.

Discussion
- What landscape types can you find in North America?
- Using the map to help you, discuss where each of the photographs on pages 50–51 was taken.
- Which landscape would you most like to visit?

2 ▼ Huge cactus plants grow in the Arizona Desert in the US.

1 ▼ The Rocky Mountains stretch from Canada to Mexico.

3 ▼ The Maya built this pyramid in the Mexican rainforest 3000 years ago.

Unit 9 — North America

Mapwork
Using an atlas, find out the names of ten different states in the US.

Key words
- cactus
- coniferous forest
- rainforest
- resources

Key
- Mountain
- Desert
- Grassland
- Northern forests
- Rainforest
- Ice cap

▼ New York has many skyscrapers.

▼ In Greenland people can only live around the coast.

Investigation
Collect your own photographs of the US for a class display.

Unit 9 North America

Lesson 2: Finding out about Canada

What is Canada like?

Canada is the second largest country in the world. It stretches over 5000 km across North America from the Atlantic Ocean to the Pacific Ocean. Canada is also a country of vast landscapes. Swamps, lakes and forests cover around half the country. In the west, the Rocky Mountains rise high into the sky.

▲ Inside the Arctic Circle the northern lights flicker across the night sky. These brilliant natural light shows attract many tourists.

Discussion
- How large is Canada?
- Why is Canada divided into time zones?
- What is remarkable about Canada?

Key words
border
First Nations
fossil fuels
maple leaf
northern lights

Data bank
Area 10 million sq km²
Highest mountain Mount Logan 5959 m
Longest river The Mackenzie River 4241 km
Coastline Longest coastline in the world 200 000 km
Capital city Ottawa
Some other cities Toronto, Montréal
Population 41 million

The maple leaf in the centre of the Canadian flag represents nature. The flag was designed in 1965.

Climate
Most parts of Canada are extremely cold in winter.

People
The First Nations were over-run by European settlers. The French came first, then the British.

Resources
Canada is a major producer of oil and gas. It is also important for wheat and timber.

Environment
Pine beetles have killed many trees. In Alberta the land has been damaged by pollution and fracking.

Unit 9 North America

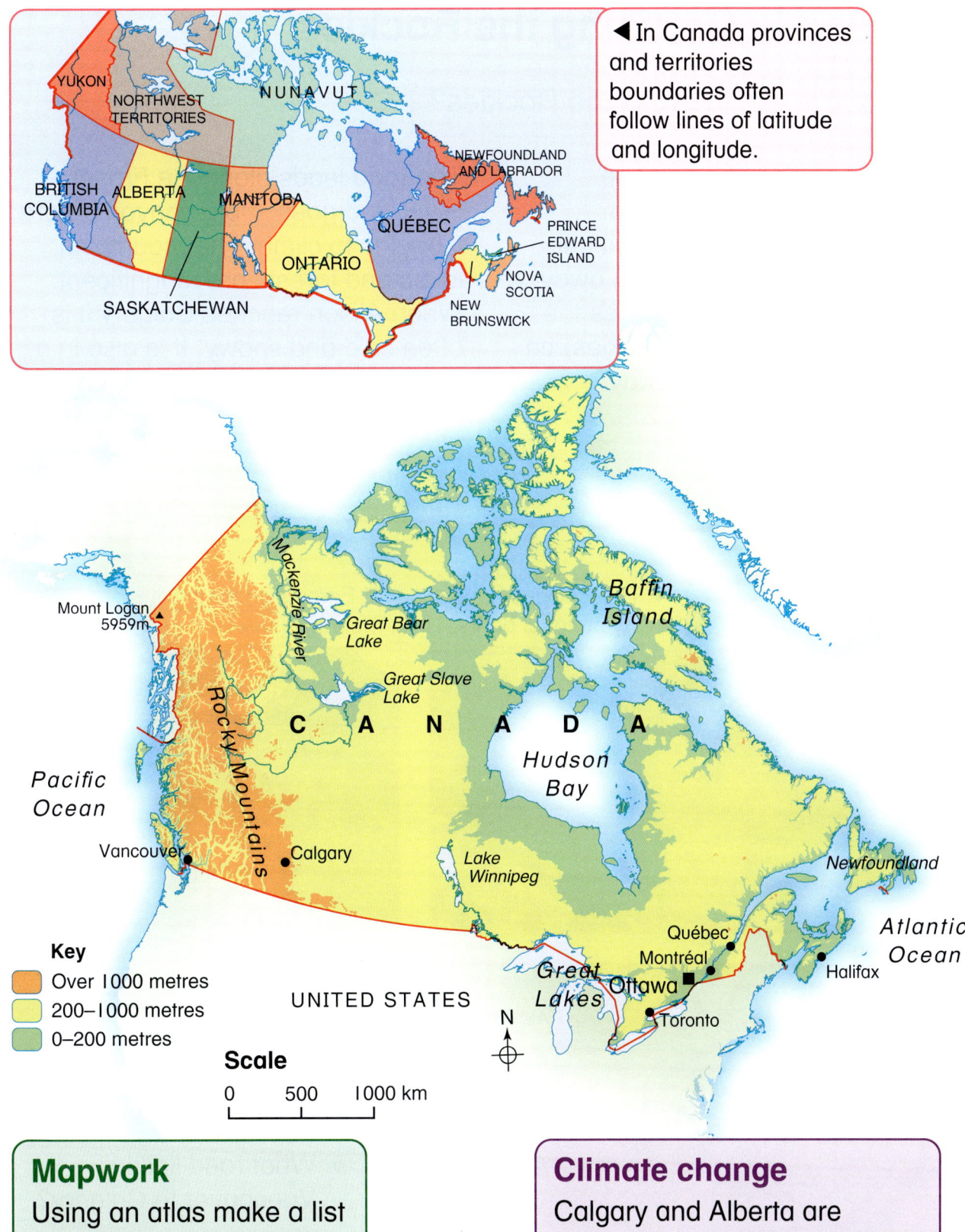

◀ In Canada provinces and territories boundaries often follow lines of latitude and longitude.

Key
- Over 1000 metres
- 200–1000 metres
- 0–200 metres

Scale
0   500   1000 km

Mapwork
Using an atlas make a list of countries like Canada which only have a border with one other country.

Climate change
Calgary and Alberta are famous for producing bitumen. This is a source of fossil fuel. Find out what fossil fuels are.

Unit 9 North America

Lesson 3: Crossing the Rockies

What is it like to cross the Rockies?

Alison lives in Vancouver on the Pacific coast of Canada. Sometimes she goes to visit her brother in Calgary 1000 km away. The journey takes her across the Rocky Mountains (the Rockies) on the Trans-Canada Highway.

The road leads along the Fraser River and goes into a canyon before it begins to climb. The mountains, lakes and forests are magnificent. When Alison reaches Calgary it is often cold and snowy. It is also in a different time zone.

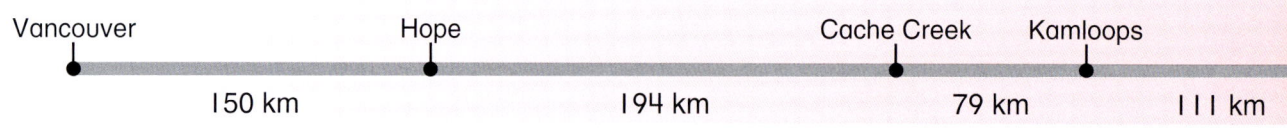

Vancouver — 150 km — Hope — 194 km — Cache Creek — 79 km — Kamloops — 111 km

▼ The Rockies have many national parks.

▼ Forests cover the mountain slopes.

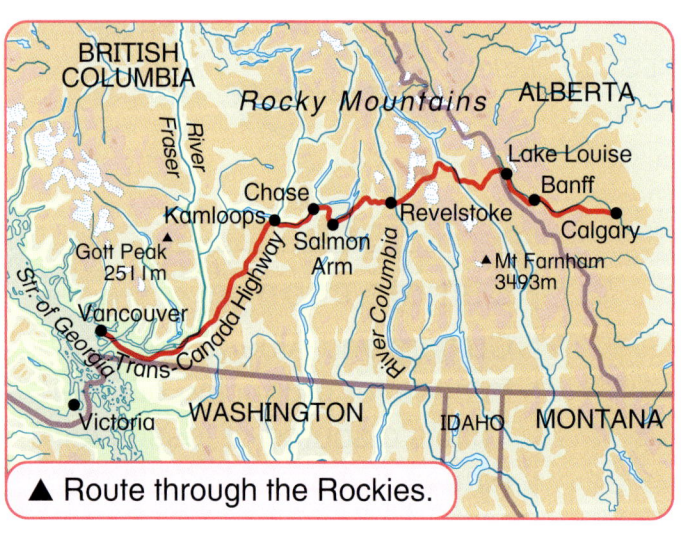

▲ Route through the Rockies.

Discussion
- What road links Vancouver to Calgary?
- What do you think you would like most about Alison's journey?

Unit 9   North America

Key words
- canyon
- mountain pass
- time zone

Alison enjoys being in the mountains. In the winter she goes skiing and in summer she goes hiking. Alison is also interested in the plants and wildlife. Once she even saw a bear in the distance.

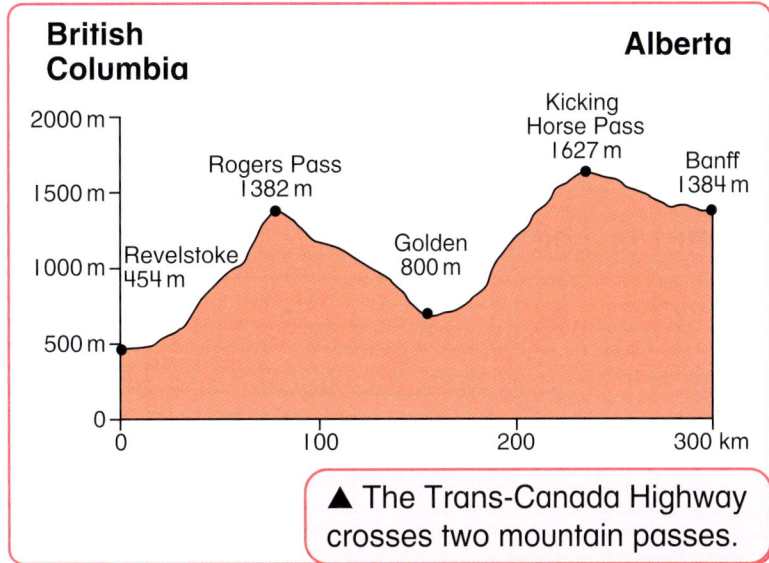

▲ The Trans-Canada Highway crosses two mountain passes.

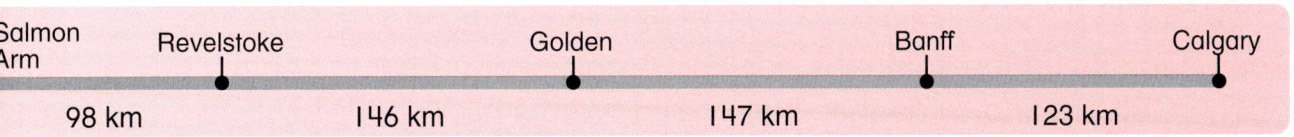

▼ People enjoy hiking in the summer.

▲ Bear, elk and other wild animals are quite common.

Investigation
Make a slideshow presentation with six slides about the Trans-Canada Highway and the Rocky Mountains.

Mapwork
Make your own diagram map of the Trans-Canada Highway showing places along the route from the Pacific to Atlantic Ocean coasts.

Summary
In this unit you have learnt:
- about the landscape of North America
- what makes Canada special
- about the Rocky Mountains.

Unit 10   Asia

Lesson 1: The Gulf

What is The Gulf?

Key words		
Arab	culture	states
desert	oil reserves	

The Gulf is the sea which separates Saudi Arabia and Iran. It is about 1000 kilometres long. The Gulf is one of the most important regions in the world for oil and gas. This has made the countries around The Gulf extremely rich.

The countries on the southern side of The Gulf are known as the 'Gulf states'. Saudi Arabia and Iraq are the largest of the Gulf states. Some of the others are very small.

The Gulf states are all Arab countries. Traditions and culture are very important. Islam is the main religion. At school children learn Arabic script. Friday, being a day of prayer, is not a workday.

◀ Long robes are widely worn in ▶ The Gulf. Men wear a white kandura and women a black abaya.

Unit 10 Asia

Discussion
- What is The Gulf?
- What makes The Gulf important?
- What else is special about The Gulf?

Data bank
- Some of the oldest civilisations in the world developed in The Gulf states.
- The Gulf has around half the world's oil reserves.
- The Gulf is home to around 700 species of fish.

▼ Dates grow well in the desert climate.

▼ The Gulf was once a centre for fishing and pearl diving.

▼ Oil was first discovered in The Gulf in the 1950s.

Mapwork
Working from an atlas, write the names of The Gulf states in alphabetical order.

Investigation
What can you find in and around your classroom which is made of oil? Make a list.

Unit 10 Asia

Lesson 2: Learning about the United Arab Emirates

UAE

What is the United Arab Emirates like?

The United Arab Emirates or UAE is a single country made from a group of seven states (emirates) on the southern shore of The Gulf. Since oil was first discovered in the 1950s the UAE has been almost completely transformed.

▶ The Burj Khalifa skyscraper is the tallest building in the world (828 metres).

Discussion
- How many states are there in the UAE?
- What traditions are maintained?
- How do you think the discovery of oil transformed the UAE?

Key words
emirate
mosque
nomadic
state
theme park
villa

Unit 10 Asia

▶ Camel racing is popular in the UAE. It is linked to the traditions of the nomadic people who live in the desert.

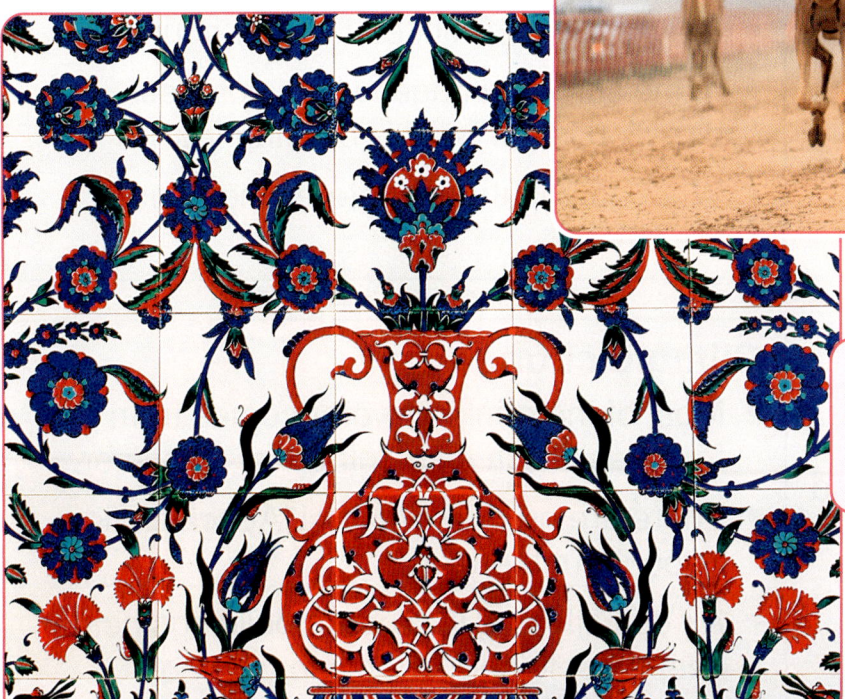

◀ Many works of art have been inspired by religion. These tiles are in a mosque in Abu Dhabi.

▲ The colours in the UAE flag symbolise Arab unity.

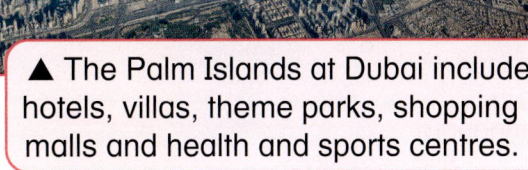

▲ The Palm Islands at Dubai include hotels, villas, theme parks, shopping malls and health and sports centres.

Mapwork
Working from an atlas, find three or more mountains around the world which are similar in height to the Burj Khalifa skyscraper.

Investigation
Research how the palm islands were created and find some photographs of the different palm islands to make a display.

Unit 10 Asia

Lesson 3: Exploring the United Arab Emirates

Key words

environmental footprint	oasis
business centre	sandstorm
	satellite image

The UAE today

The population of the UAE has multiplied eight times since 1975. Abu Dhabi is the capital and business centre. Dubai is another important city. There are oases in the desert areas.

Temperatures can be as low as 10 °C in winter and over 40 °C in summer. Sometimes strong winds cause sandstorms. Modern buildings help people live in these difficult conditions.

Discussion
- How is the UAE linked to other countries?
- In the satellite image, which compass direction is the wind is blowing from?
- What do you think is interesting about the UAE?

Climate change
- Most of the drinking water comes from factories which take salt out of seawater.
- There are no rivers in the UAE which flow throughout the year.
- Air conditioning in modern buildings helps keep people comfortable in very hot weather. However this uses a lot of energy, which affects the environmental footprint.

Links with other countries

Trade

The UAE sells oil to many countries in Asia. This diagram shows where the oil goes.

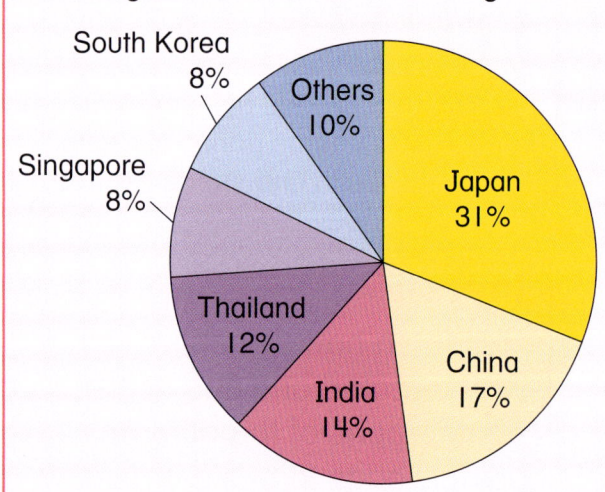

Transport

Emirates and Etihad airlines are both based in the UAE and have flights to cities across six continents.

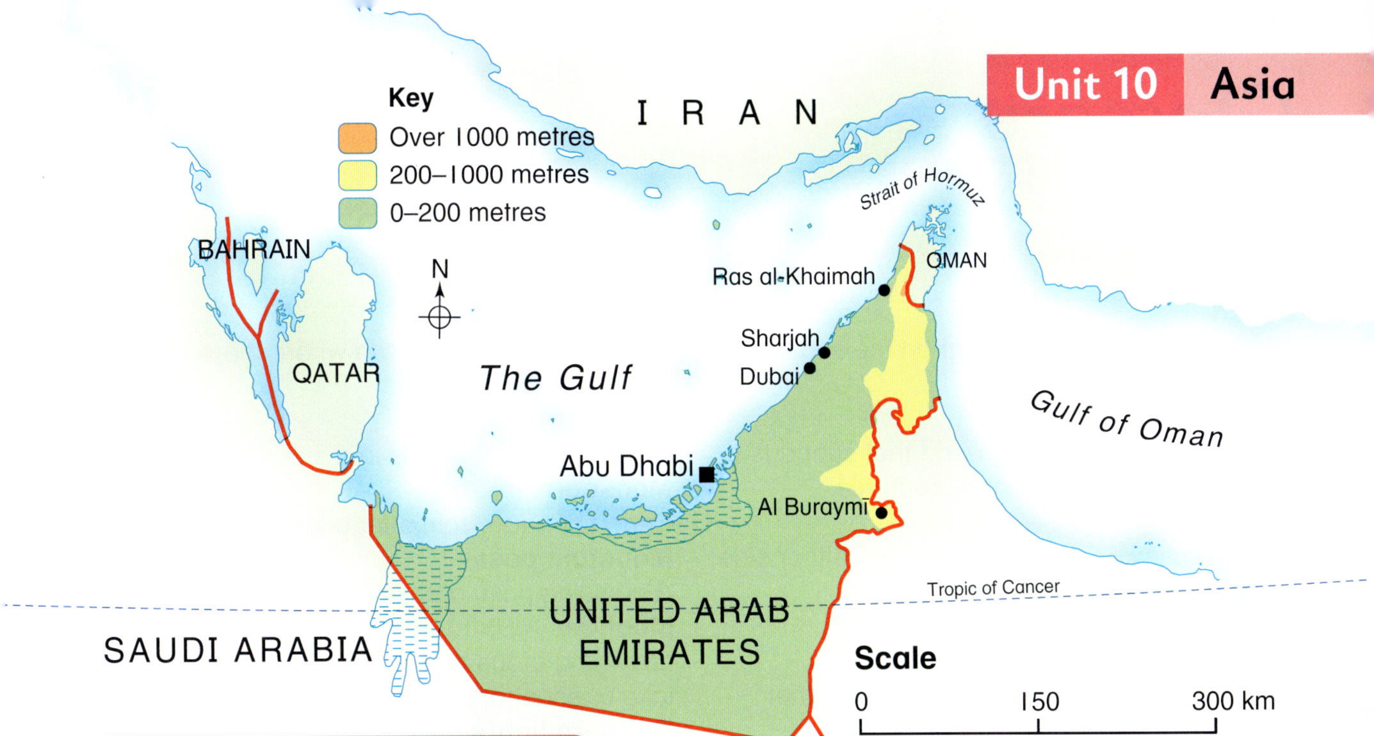

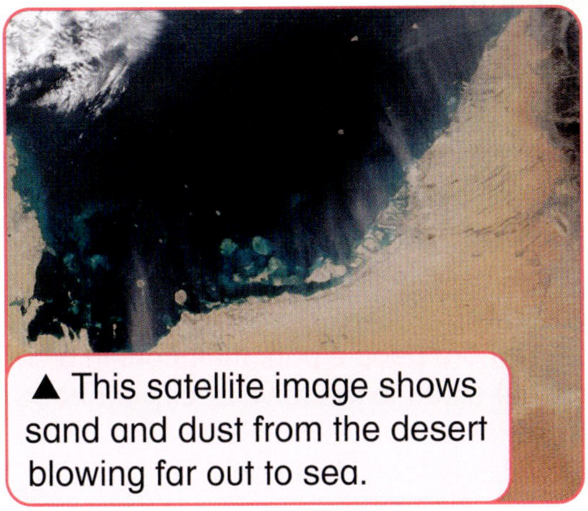

Unit 10 Asia

Key
- Over 1000 metres
- 200–1000 metres
- 0–200 metres

IRAN, BAHRAIN, QATAR, The Gulf, Strait of Hormuz, Ras al-Khaimah, OMAN, Sharjah, Dubai, Gulf of Oman, Abu Dhabi, Al Buraymi, Tropic of Cancer, UNITED ARAB EMIRATES, SAUDI ARABIA, OMAN

Scale: 0, 150, 300 km

▲ This satellite image shows sand and dust from the desert blowing far out to sea.

Sport
The Emirates Stadium in London is home to the Arsenal football team and one of the largest stadiums in the UK. The UAE also sponsors other sports such as motor racing.

Investigation
Write a few sentences about environmental footprints and what they tell us.

Mapwork
Using an outline map of Asia draw arrows to show the shipping routes from the UAE to the countries it supplies with oil.

Summary
In this unit you have learnt:
- about the Gulf states
- about the UAE
- how the UAE is changing.

61

Glossary

Beach
The material which builds up on the seashore and is washed by the tide.
Bog
An area of wet, muddy ground that supports a special range of plants and creatures.
Canyon
A deep valley or gorge with cliffs on either side.
Coast
Where the land meets the sea.
Culture
The music, art, language, beliefs and other traditions of a group of people.
Dam
A barrier across a river which creates a lake or reservoir.
Environmental footprint
The resources (food, clothing, water and so on) which a person uses in their daily life.
Estuary
Where a river meets the sea and the water becomes salty.
Fairtrade
System for buying goods which gives farmers and other workers a fair price for their products.
First Nations
A term used to describe the people who lived in Canada before Europeans arrived.
Food miles
The distance from where food is grown to the person who uses it.
Fossil fuels
Fuels like oil and gas that create carbon emissions.
Grid square
System of squares for identifying places on a map.
Habitat
The place where a community of plants and animals live.
Headland
A narrow piece of land that juts out into the sea.

High street
The main street in a town where there are shops.
Meander
A series of wide bends in a stream or river.
Monsoon
Period of heavy rain after a long dry period.
Mountain pass
A crossing point across a range of mountains.
Mudflats
Low-lying areas of mud found at the edge of a lake or seashore.
Northern lights
Colourful lights in the night sky seen in polar regions, caused by particles from the sun.
Oasis
An isolated area in the desert where water comes to the surface.
Oil reserves
The amount of oil in the world.
Park and ride
A scheme where people park their cars in a car park and get a bus to the town or city centre.
Pedestrianised
An area which is for walking only, rather than travelling in a vehicle.
Picture scales
A set of drawings to show changes on a theme or topic.
Pumping station
Place where water is drawn out of a river or reservoir.
Resources
Metals, crops and other substances which people find useful.
Shingle
Small to medium-sized pebbles which are found on a beach.
Tide
The daily change of level in seawater caused by the action of the sun and the moon.
Time zone
An area where all places share the same time, usually about 15 degrees of longitude apart.

Index

Abu Dhabi 59, 60
advertisements 36, 37
Africa 10, 11, 15
Amazon, river 10
Arctic Circle 52
Arizona Desert 50
Aswan High Dam, Egypt 11
Asia 56–61
Atlantic Ocean 39, 52
Australia 15

Bann, river 38, 39
Belfast, Northern Ireland 38, 39
Berlin, Germany 44, 45
blizzards 14
buildings 20, 22, 32, 33, 36, 37, 47, 58, 60

Canada 4, 15, 50, 52–55
Cancun, Mexico 23
caves 5, 7
Chang Jiang (Yangtze), river 10
cliffs 3, 4, 5, 6, 7
climate 52, 57
climate change 2, 15, 37, 41, 44, 53, 61
clouds 16, 17, 18, 19
coasts 2–7, 52
Cornwall 3
crops 11, 16, 22, 26

Danube, river 44
Djenné, Mali 22
Dinslaken, Germany 46–49
Dover, UK 23
drought 15
Dubai 59, 60
dunes 4

East Kilbride, Scotland 20–21
Egypt 10, 11
Elbe, river 44
Environment Agency 12
estuary 8, 9

factories 22, 23, 39, 43, 47, 49, 61
farming 11, 12, 16, 26, 27, 28, 34, 39, 40, 41
ferries 39, 46
First Nations 52
floods 11, 14, 15, 44
food 24, 26–31, 45
fossil fuels 53
Foyle, river 43

gas 52, 56
Germany 15, 44–49
Giant's Causeway, Northern Ireland 38
Glasgow, Scotland 34
gorge 9, 44
Greenland 51
Gulf, The 56, 57

habitats 7, 43
headlands 4, 5
Highlands, the, Scotland 3, 14
high-speed railway 45
hurricanes 14

improvement schemes 35
Indian Ocean 19
Iraq 56
irrigation 11
Isle of Wight 3
Istanbul, Turkey 22

Leigh-on-Sea, England 3
Londonderry (Derry), Northern Ireland 39, 40–43
Lowell, US 23

maps 2, 6, 10, 16, 21, 26, 38, 41, 45, 46, 48, 51, 53, 54, 56, 61
markets 22, 26, 47
meanders 9
Mediterranean Sea 11
Met Office 17, 19
Mexico 23, 50, 51
Mississippi–Missouri, river 10
motorways 39, 45, 46
mudflats 2, 3
Murray Darling, river 10

Needles, The, England 3
New York, US 51
Nile, river 10–11
North America 50–55
Northern Ireland 2, 38–43
northern lights 52

Oder, river 44
oil 52, 56, 57, 58, 60

Pacific Ocean 52
park and ride 35
pedestrianisation 35, 48
ports 23, 43

rain 14, 15, 16, 17, 18, 19, 26, 39, 40
Rhine, river 44
Ruhr, Germany 46, 47
rivers 8–13, 15, 39, 43, 44, 61
rock stack 5
Rocky Mountains 50, 51, 52, 54, 55

Sahara Desert 10, 11, 22
salmon 9
Scotland 3, 14, 20, 21, 26, 38
sea 4, 5
seashore 2, 3, 7
seaside towns and resorts 6, 23
services 24, 25
shops 13, 20, 21, 22, 26–31, 33, 43, 47, 59
Sidmouth, England 6
snow 14, 18, 39, 54
sport 61
storms 14, 15, 60
streams 8, 9
Sudan 11
Sumida, river 13

Tees, river 12
temperature 15, 16, 17, 18, 19, 60
Thames, river 12, 28
tide 7
tourists 11, 23, 52
towns 6, 20–25, 32–37
town planners 20, 34
trade 22, 60
traffic calming 35
transport 20, 24, 29, 39, 45, 60
tributary 8

United Arab Emirates 58–61
United Kingdom 2, 5, 38–43
United States 19, 23, 50, 51

waterfall 9
waves 4, 6
weather balloons 17
weather forecasts 16–17
weather patterns 14–19
weather satellites 17
Weser, river 44
windmills 33
winds 4, 14, 15, 17, 18, 39, 44, 60
work 12, 39, 44

63

William Collins' dream of knowledge for all began with the publication of his first book in 1819.

A self-educated mill worker, he not only enriched millions of lives, but also founded a flourishing publishing house. Today, staying true to this spirit, Collins books are packed with inspiration, innovation and practical expertise.
They place you at the centre of a world of possibility and give you exactly what you need to explore it.

Published by Collins
An imprint of HarperCollins*Publishers*
The News Building, 1 London Bridge Street, London,
SE1 9GF, UK

HarperCollins*Publishers*
Macken House, 39/40 Mayor Street Upper, Dublin 1,
D01 C9W8, Ireland

Browse the complete Collins catalogue at
collins.co.uk

© HarperCollinsPublishers Limited 2025

Maps © Collins Bartholomew 2025

10 9 8 7 6 5 4 3 2

ISBN 978-0-00-872831-1

All rights reserved. No part of this publication may be reproduced, stored in a retrieval system, or transmitted in any form by any means, electronic, mechanical, photocopying, recording or otherwise, without the prior written permission of the Publisher or a licence permitting restricted copying in the United Kingdom issued by the Copyright Licensing Agency Ltd, 5th Floor, Shackleton House, 4 Battle Bridge Lane, London SE1 2HX.

Without limiting the exclusive rights of any author, contributor or the publisher, any unauthorised use of this publication to train generative artificial intelligence (AI) technologies is expressly prohibited. HarperCollins also exercise their rights under Article 4(3) of the Digital Single Market Directive 2019/790 and expressly reserve this publication from the text and data mining exception.

British Library Cataloguing-in-Publication Data

A catalogue record for this publication is available from the British Library.

Authors: Stephen Scoffham and Colin Bridge (with additional original input from by Terry Jewson)
Publisher: Laura White
Product manager: Natasha Paul
Development editor: Judith Walters
Copyeditor and proofreader: Catherine Dakin
Cover designer and illustrator: Steve Evans
Internal illustrators: Jouve India Private Ltd and Hannah Drennan, Beehive Illustration
Typesetter: David Jimenez
Production controller: Alhady Ali
Printed and bound in Great Britain by Bell and Bain Ltd, Glasgow

This book is produced from independently certified FSC™ paper to ensure responsible forest management.

For more information visit: www.harpercollins.co.uk/green
collins.co.uk/sustainability

Acknowledgements

The publishers gratefully acknowledge the permission granted to reproduce the copyright material in this book. Every effort has been made to trace copyright holders and to obtain their permission for the use of copyright material. The publishers will gladly receive any information enabling them to rectify any error or omission at the first opportunity.

P18tr, P24tr, P25tc © Stephen Scoffham; P19r 'Example weather diary', Met Office, © Crown copyright, 2009. Open Government Licence for Public Sector Information v3.0; P21t © Aerial Photography Solutions Ltd; P34r PA Images/Alamy Stock Photo; P41l © Richard Webb/Geograph; P49bl Olaf Doering/Alamy Stock Photo; P56 Mahmoud Rahall/Alamy Stock Photo.

All other photos Shutterstock.